Natural Gas Future

Natural Gas Future

A World Without Oil

Richard L. Itteilag

authorHOUSE®

AuthorHouse™
1663 Liberty Drive
Bloomington, IN 47403
www.authorhouse.com
Phone: 1-800-839-8640

© 2012 by Richard L. Itteilag. All rights reserved.

No part of this book may be reproduced, stored in a retrieval system, or transmitted by any means without the written permission of the author.

Published by AuthorHouse   08/21/2012

ISBN: 978-1-4772-6377-8 (sc)
ISBN: 978-1-4772-6376-1 (e)

Library of Congress Control Number: 2012915570

Any people depicted in stock imagery provided by Thinkstock are models, and such images are being used for illustrative purposes only.
Certain stock imagery © Thinkstock.

This book is printed on acid-free paper.

Because of the dynamic nature of the Internet, any web addresses or links contained in this book may have changed since publication and may no longer be valid. The views expressed in this work are solely those of the author and do not necessarily reflect the views of the publisher, and the publisher hereby disclaims any responsibility for them.

Contents

The U.S. as the Saudi Arabia of Natural Gas 1
Advances in the Exploration and Production Sector 4
Coiled Tubing 7
Measurement While Drilling 8
Slimhole Drilling 9
Offshore Production 10
Liquefied Natural Gas 11
LNG Delivery Facility with Tanker 12
Extensive Pipeline Network 13
The Natural Gas Gathering System 15
Natural Gas Fuel Cells 17
Clean Electricity 18
How a Fuel Cell Works 19
Coiled Tubing 23
Liquefied Natural Gas 25
Natural Gas Fuel Cells 26
Clean Electricity 27
How a Fuel Cell Works 28
Transportation Process and Flow 30
The Natural Gas Gathering System 32
The Natural Gas Gathering System 37
The Natural Gas Processing Plant 44
The Transmission Grid and Compressor Stations 45
Natural Gas Market Centers/Hubs 47

The U.S. as the Saudi Arabia of Natural Gas

Natural gas is a vital component of the world's supply of energy. It is one of the cleanest, safest and most useful of all energy sources. Despite its importance, however, there are many misconceptions about natural gas. For instance, the word 'gas' itself has a variety of different uses, and meanings. When we fuel our car, we put 'gas' in it. However, the gasoline that goes into your vehicle, while a fossil fuel itself, is very different from natural gas. The 'gas' in the common barbecue is actually propane, which, while closely associated and commonly found in natural gas, is not really natural gas itself.

While commonly grouped in with other fossil fuels and sources of energy, there are many characteristics of natural gas that make it unique. Below is a bit of background information about natural gas, what it is exactly, how it is formed, and how it is found in nature.Natural gas, in itself, might be considered an uninteresting gas—it is colorless, shapeless, and odorless in its pure form. Quite uninteresting—except that natural gas is combustible, abundant in the United States and, when burned, it gives off a great deal of energy and few harmful emissions. Unlike other fossil fuels, natural gas is clean burning and emits lower levels of potentially harmful byproducts into the air.

We require energy constantly, to heat and cool our homes, cook our food, and generate our electricity. It is this need for energy that has elevated natural gas to such a level of importance in our society, and in our lives.Natural gas is a combustible mixture of hydrocarbon gases. While natural gas is formed primarily of methane, it can also include ethane, propane, butane and pentane. The composition of natural gas can vary widely.

In its purest form, such as the natural gas that is delivered to your home, it is almost pure methane. Methane is a molecule made up of one carbon atom and four hydrogen atoms, and is referred to as CH4. The distinctive "rotten egg" smell that we often associate with natural gas is actually an odorant called mercaptan that is added to the gas before it is delivered to the end-user. Mercaptan aids in detecting any leaks.

Natural gas is considered 'dry' when it is almost pure methane, having had most of the other commonly associated hydrocarbons removed. When other hydrocarbons are present, the natural gas is 'wet'.

Natural gas is a vital component of the world's supply of energy. It is one of the cleanest, safest and most useful of all energy sources. Despite its importance, however, there are many misconceptions about natural gas. For instance, the word 'gas' itself has a variety of different uses, and meanings. When we fuel our car, we put 'gas' in it. However, the gasoline that goes into your vehicle, while a fossil fuel itself, is very different from natural gas. The 'gas' in the common barbecue is actually propane, which, while closely associated and commonly found in natural gas, is not really natural gas itself. While commonly grouped in with other fossil fuels and sources of energy, there are many characteristics of natural gas that make it unique. Below is a bit of background information about natural gas, what it is exactly, how it is formed, and how it is found in nature.

Natural gas is a vital component of the world's supply of energy. It is one of the cleanest, safest, and most useful of all energy sources. Despite its importance, however, there are many misconceptions about natural gas. For instance, the word 'gas' itself has a variety of different uses, and meanings. When we fuel our car, we put 'gas' in it. However, the gasoline that goes into your vehicle, while a fossil fuel itself, is very different from natural gas. The 'gas' in the common barbecue is actually propane, which, while closely associated and commonly found in natural gas, is not really natural gas itself. While commonly grouped in with other fossil fuels and sources of energy, there are many characteristics of natural gas that make it unique. Below is a bit of background information about natural gas, what it is exactly, how it is formed, and how it is found in nature.

The preservation of our environment is a very important and pressing topic, particularly when dealing with energy issues. The advancement of technology, particularly technologies that allow the cleaner use of fossil fuels, may provide many environmental benefits and allow us to use cleaner energy. This section focuses on the environmental issues related to the use of natural gas, as well as the advancement of new and exciting technologies within the industry. Beginning in the 1990s, more power generators began to use natural gas to make electricity, causing demand for natural gas to grow substantially. The natural gas industry has been able to keep pace with growing demand and produce greater amounts of natural gas through technological innovation. These innovations have enabled the development of natural gas from shale and other formerly "unconventional" formations that are found in abundance across the United States, as well as development from traditional offshore and onshore formations. Below is a brief list of some of the major technological advancements that have been made recently:

Technological innovation in the exploration and production sector has equipped the industry with the equipment and practices necessary to continually increase the production of natural gas to meet rising demand. These technologies serve to make the exploration and production of natural gas more efficient, safe, and environmentally friendly. Even as natural gas deposits are increasingly produced from "unconventional" formations such as shale rock, the exploration and production industry has not only kept up its production pace, but in fact has improved the general nature of its operations, contributing to an unprecedented 39 percent increase in the size of U.S. resources since 2006.

Over the past few decades the oil and natural gas industry has transformed into one of the most technologically advanced industries in the United States. New innovations have reshaped the industry into a technology leader. This section will discuss the role of technology in the evolution of the natural gas industry, focusing on technologies in the exploration and production sector, as well as a few select innovations that have had a profound effect on the potential for natural gas.

Advances in the Exploration and Production Sector

Liquefied Natural Gas
Natural Gas Fuel Cells
Natural Gas Technology Resources

Beginning in the 1990s, more power generators began to use natural gas to make electricity, causing demand for natural gas to grow substantially. The natural gas industry has been able to keep pace with growing demand and produce greater amounts of natural gas through technological innovation. These innovations have enabled the development of natural gas from shale and other formerly "unconventional" formations that are found in abundance across the United States, as well as development from traditional offshore and onshore formations. Below is a brief list of some of the major technological advancements that have been made recently:

Advances in the Exploration and Production Sector

Technological innovation in the exploration and production sector has equipped the industry with the equipment and practices necessary to continually increase the production of natural gas to meet rising demand. These technologies serve to make the exploration and production of natural gas more efficient, safe, and environmentally friendly. Even as natural gas deposits are increasingly produced from "unconventional" formations such as shale rock, the exploration and production industry has not only kept up its production pace, but in fact has improved the general nature of its operations, contributing to an unprecedented 39 percent increase in the size of U.S. resources since 2006.

According to a Department of Energy Report, "Environmental Benefits of Advanced Oil and Gas Exploration and Production Technology," released in 1999 and still one of the most indepth analyses available as of 2010:

22,000 fewer wells are needed on an annual basis to develop the same amount of oil and gas reserves as were developed in 1985.

Had technology remained constant since 1985, it would take two wells to produce the same amount of oil and natural gas as one 1985 well. However, advances in technology mean that one well today can produce two times as much as a single 1985 well.

Drilling wastes have decreased by as much as 148 million barrels due to increased well productivity and fewer wells.

The drilling footprint of well pads has decreased by as much as 70 percent due to advanced drilling technology, which is extremely useful for drilling in sensitive areas.

By using modular drilling rigs and slimhole drilling, the size and weight of drilling rigs can be reduced by up to 75 percent over traditional drilling rigs, reducing their surface impact.

Had technology, and thus drilling footprints, remained at 1985 levels, today's drilling footprints would take up an additional 17,000 acres of land.

New exploration techniques and vibrational sources mean less reliance on explosives, reducing the impact of exploration on the environment.

New exploration techniques and vibrational sources mean less reliance on explosives, reducing the impact of exploration on the environment. New exploration techniques and vibrational sources mean less reliance on explosives, reducing the impact of exploration on the environment.

Some of the major recent technological innovations in the exploration and production sector include:

Richard L. Itteilag

Advanced 3-D Seismic Imaging and 4-D Seismic Imaging

The development of seismic imaging in three dimensions greatly changed the nature of natural gas exploration. This technology uses traditional seismic imaging techniques, combined with powerful computers and processors, to create a three-dimensional model of the subsurface layers. 4-D seismology expands on this, by adding time as a dimension, allowing exploration teams to observe how subsurface characteristics change over time. Exploration teams can now identify natural gas prospects more easily, place wells more effectively, reduce the number of dry holes drilled, reduce drilling costs, and cut exploration time. This leads to both economic and environmental benefits.

CO2-Sand Fracturing—Fracturing techniques have been used since the 1970s to help increase the flow rate of natural gas and oil from underground formations. CO2-Sand fracturing involves using a mixture of sand proppants and liquid CO2 to fracture formations, creating and enlarging cracks through which oil and natural gas may flow more freely. The CO2 then vaporizes, leaving only sand in the formation, holding the newly enlarged cracks open. Because there are no other substances used in this type of fracturing, there are no 'leftovers' from the fracturing process that must be removed. This means that, while this type of fracturing effectively opens the formation and allows for increased recovery of oil and natural gas, it does not damage the deposit, generates no below ground wastes, and protects groundwater resources.

Coiled Tubing

Coiled tubing technologies replace the traditional rigid, jointed drill pipe with a long, flexible coiled pipe string. This greatly reduces the cost of drilling, as well as providing a smaller drilling footprint, requiring less drilling mud, faster rig set up, and reducing the time normally needed to make drill pipe connections. Coiled tubing can also be used in combination with slimhole drilling to provide very economic drilling conditions, and less impact on the environment.

Measurement While Drilling

Measurement-While-Drilling (MWD) systems allow for the collection of data from the bottom of a well as it is being drilled. This allows engineers and drilling teams access to up-to-the-second information on the exact nature of the rock formations being encountered by the drill bit. This improves drilling efficiency and accuracy in the drilling process, allows better formation evaluation as the drill bit encounters the underground formation, and reduces the chance of formation damage and blowouts.

Slimhole Drilling

Slimhole drilling is exactly as it sounds; drilling a slimmer hole in the ground to get to natural gas and oil deposits. In order to be considered slimhole drilling, at least 90 percent of a well must be drilled with a drill bit less than six inches in diameter (whereas conventional wells typically use drill bits as large as 12.25 inches in diameter). Slimhole drilling can significantly improve the efficiency of drilling operations, as well as decrease its environmental impact. In fact, shorter drilling times and smaller drilling crews can translate into a 50 percent reduction in drilling costs, while reducing the drilling footprint by as much as 75 percent. Because of its low cost profile and reduced environmental impact, slimhole drilling provides a method of economically drilling exploratory wells in new areas, drilling deeper wells in existing fields, and providing an efficient means for extracting more natural gas and oil from un-depleted fields.

Offshore Production

The offshore oil and natural gas production sector is sometimes compared to the aeronautics field and NASA due to achievements in deepwater drilling that have been facilitated by state of the art technology. New technology, including improved offshore drilling rigs, dynamic positioning devices and sophisticated navigation systems are allowing safe, efficient offshore drilling in waters more than 10,000 feet deep. Visit the offshore drilling section to learn more.

Hydraulic Fracturing also called "Fracking," or "Frac'ing"—Used to free natural gas that is trapped in shale rock formations. A liquid mix that is 99 percent water and sand is injected into the rock at very high pressure, creating fractures within the rock that provide the natural gas a path to flow to the wellhead. The fracking fluid mix also helps to keep the formation more porous. Hydraulic fracturing is now widely used, with more than 90 percent of the natural gas wells in the United States having used it to boost production at some time.

The above technological advancements provide only a snapshot of the increasingly sophisticated technology being developed and put into practice in the exploration and production of natural gas and oil. New technologies and applications are being developed constantly, and serve to improve the economics of producing natural gas, allow for the production of deposits formerly considered too unconventional or uneconomic to develop, and ensure that the supply of natural gas keeps up with steadily increasing demand. Sufficient domestic natural gas resources exist to help fuel the U.S. for a significant period of time, and technology is playing a huge role in providing low-cost, environmentally sound methods of extracting these resources.

Two other technologies that are revolutionizing the natural gas industry include the increased use of liquefied natural gas, and natural gas fuel cells. These technologies are discussed below:

Liquefied Natural Gas

Cooling natural gas to about -260°F at normal pressure results in the condensation of the gas into liquid form, known as Liquefied Natural Gas (LNG). LNG can be very useful, particularly for the transportation of natural gas, since LNG takes up about one six hundredth the volume of gaseous natural gas. Advances in technology are reducing the costs associated with the liquification and regasification of LNG. Because it is easy to transport, LNG can serve to make economical stranded natural gas deposits from around the globe for which the construction of pipelines is uneconomical.

LNG Delivery Facility with Tanker

Source

LNG, when vaporized to gaseous form, will only burn in concentrations of between 5 and 15 percent mixed with air. In addition, LNG, or any vapor associated with LNG, will not explode in an unconfined environment. Thus, in the unlikely event of an LNG spill, the natural gas has little chance of igniting an explosion. Liquification removes oxygen, carbon dioxide, sulfur, and water from the natural gas, resulting in LNG that is almost pure methane.

LNG is typically transported by specialized tanker with insulated walls, and is kept in liquid form by autorefrigeration, a process in which the LNG is kept at its boiling point, so that any heat additions are countered by the energy lost from LNG vapor that is vented out of storage and used to power the vessel.

The increased use of LNG is allowing for the production and marketing of natural gas deposits that were previously economically unrecoverable. Although it currently accounts for only about 1 percent of natural gas used in the United States, it is expected that LNG imports will provide a steady, dependable source of natural gas for U.S. consumption.

Extensive Pipeline Network

Transportation Process and Flow

Overview | Gathering System | Processing Plant | Transmission Grid | Market Centers/Hubs | Underground Storage | Peak Shaving

Overview

Transporting natural gas from the wellhead to the final customer involves several physical transfers of custody and multiple processing steps. A natural gas pipeline system begins at the natural gas producing well or field. Once the gas leaves the producing well, a pipeline gathering system directs the flow either to a natural gas processing plant or directly to the mainline transmission grid, depending upon the initial quality of the wellhead product.

The processing plant produces pipeline-quality natural gas. This gas is then transported by pipeline to consumers or is put into underground storage for future use. Storage helps to maintain pipeline system operational integrity and/or to meet customer requirements during peak-usage periods.

Transporting natural gas from wellhead to market involves a series of processes and an array of physical facilities. Among these are:

Gathering Lines—These small-diameter pipelines move natural gas from the wellhead to the natural gas processing plant or to an interconnection with a larger mainline pipeline.

Processing Plant—This operation extracts natural gas liquids and impurities from the natural gas stream.

Mainline Ttransmission Systems—These wide-diameter, long-distance pipelines transport natural gas from the producing area to market areas.

Market Hubs/Centers—Locations where pipelines intersect and flows are transferred.

Underground Storage Facilities—Natural gas is stored in depleted oil and gas reservoirs, aquifers, and salt caverns for future use.

Peak Shaving—System design methodology permitting a natural gas pipeline to meet short-term surges in customer demands with minimal infrastructure. Peaks can be handled by using gas from storage or by short-term line-packing.

The Natural Gas Gathering System

A natural gas pipeline system begins at a natural gas producing well or field. In the producing area many of the pipeline systems are primarily involved in "gathering" operations. That is, a pipeline is connected to a producing well, converging with pipes from other wells where the natural gas stream may be subjected to an extraction process to remove water and other impurities if needed. Natural gas exiting the production field is usually referred to as "wet" natural gas if it still contain significant amounts of hydrocarbon liquids and contaminants.

Under certain conditions some or all of the natural gas produced at a well may be returned to the reservoir in cycling, repressuring, or conservation operations and/or vented and flared. At this stage it is a mixture of methane and other hydrocarbons, as well as some non-hydrocarbons, existing in the gaseous phase or in a solution with crude oil. The principal hydrocarbons normally contained in the natural gas mixture are methane, ethane, propane, butane, and pentane. Typical non-hydrocarbon gases that may be present in reservoir natural gas are water vapor, carbon dioxide, helium, hydrogen sulfide, and nitrogen.

More than 210 natural gas pipeline systems. 305,000 miles of interstate and intrastate transmission pipelines.

More than 1,400 compressor stations that maintain pressure on the natural gas pipeline network and assure continuous forward movement of supplies (see map).

More than 11,000 delivery points, 5,000 receipt points, and 1,400 interconnection points that provide for the transfer of natural gas throughout the United States.

Richard L. Itteilag

24 hubs or market centers that provide additional interconnections l.e.,

400 underground natural gas storage facilities,

49 locations where natural gas can be imported/exported via pipelines, 8 LNG (liquefied natural gas) import facilities and 100 LNG peaking facilities

Natural Gas Fuel Cells

Fuel cells powered by natural gas are an extremely exciting and promising new technology for the clean and efficient generation of electricity. Fuel cells have the ability to generate electricity using electrochemical reactions as opposed to combustion of fossil fuels to generate electricity. Essentially, a fuel cell works by passing streams of fuel (usually hydrogen) and oxidants over electrodes that are separated by an electrolyte. This produces a chemical reaction that generates electricity without requiring the combustion of fuel, or the addition of heat as is common in the traditional generation of electricity. When pure hydrogen is used as fuel, and pure oxygen is used as the oxidant, the reaction that takes place within a fuel cell produces only water, heat, and electricity. In practice, fuel cells result in very low emission of harmful pollutants, and the generation of high-quality, reliable electricity. The use of natural gas-powered fuel cells has a number of benefits, including:

Clean Electricity

Fuel cells provide the cleanest method of producing electricity from fossil fuels. While a pure hydrogen, pure oxygen fuel cell produces only water, electricity, and heat, fuel cells in practice emit trace amounts of sulfur compounds and very low levels of carbon dioxide. However, the carbon dioxide produced by fuel cell use is concentrated and can be readily recaptured, as opposed to being emitted into the atmosphere.

Distributed Generation—Fuel cells can come in extremely compact sizes, allowing for their placement wherever electricity is needed. This includes residential, commercial, industrial, and even transportation settings.

Dependability—Fuel cells are completely enclosed units, with no moving parts or complicated machinery. This translates into a dependable source of electricity, capable of operating for thousands of hours. In addition, they are very quiet and safe sources of electricity. Fuel cells also do not have electricity surges, meaning they can be used where a constant, dependable source of electricity is needed.

Efficiency—Fuel cells convert the energy stored within fossil fuels into electricity much more efficiently than traditional generation of electricity using combustion. This means that less fuel is required to produce the same amount of electricity. The National Energy Technology Laboratory estimates that, used in combination with natural gas turbines, fuel cell generation facilities can be produced that will operate in the 1 to 20 Megawatt range at 70 percent efficiency, which is much higher than the efficiencies that can be reached by traditional generation methods within that output range.

This work is in the public domain in the United States because it is a work prepared by an officer or employee of the United States Government as part of that person's official duties under the terms of Title 17, Chapter 1, Section 105 of the US Code. See Copyright.

How a Fuel Cell Works

The generation of electricity has traditionally been a very polluting, inefficient process. However, with new fuel cell technology, the future of electricity generation is expected to change dramatically in the next ten to twenty years. Research and development into fuel cell technology is ongoing, to ensure that the technology is refined to a level where it is cost-effective for all varieties of electric generation requirements.

Natural Gas Technology Resources

The natural gas industry is joined by government agencies and laboratories, private research and development firms, and environmental technology groups in coming up with new technologies that may improve the efficiency, cost-effectiveness, and environmental soundness of the natural gas industry. New technology and methods emerge frequently in the natural gas industry. Below are links to a number of resources that provide information on new technological developments in the oil and natural gas industry:

Sources:

Gas Technology Institute
Department of Energy's Office of Fossil Energy
The National Energy Technology Laboratory
The Natural Gas & Oil Technology Partnership
Department of Energy's National Energy Technology
The Oil and Gas Journal

Source: ChevronTexaco Corporation

Over the past few decades the oil and natural gas industry has transformed into one of the most technologically advanced industries in the United States.

New innovations have reshaped the industry into a technology leader. This section will discuss the role of technology in the evolution of the natural gas industry, focusing on technologies in the exploration and production sector, as well as a few select innovations that have had a profound effect on the potential for natural gas. Scroll down, or click on the links below to jump ahead:

Advances in the Exploration and Production Sector

Liquefied Natural Gas
Natural Gas Fuel Cells
Natural Gas Technology Resources

Beginning in the 1990s, more power generators began to use natural gas to make electricity, causing demand for natural gas to grow substantially. The natural gas industry has been able to keep pace with growing demand and produce greater amounts of natural gas through technological innovation. These innovations have enabled the development of natural gas from shale and other formerly "unconventional" formations that are found in abundance across the United States, as well as development from traditional offshore and onshore formations. Below is a brief list of some of the major technological advancements that have been made recently:

Advances in the Exploration and Production Sector

Technological innovation in the exploration and production sector has equipped the industry with the equipment and practices necessary to continually increase the production of natural gas to meet rising demand. These technologies serve to make the exploration and production of natural gas more efficient, safe, and environmentally friendly. Even as natural gas deposits are increasingly produced from "unconventional" formations such as shale rock, the exploration and production industry has not only kept up its production pace, but in fact has improved the general nature of its operations, contributing to an unprecedented 39 percent increase in the size of U.S. resources since 2006.

According to a Department of Energy Report, "Environmental Benefits of Advanced Oil and Gas Exploration and Production Technology," released in 1999 and still one of the most indepth analyses available as of 2010:

22,000 fewer wells are needed on an annual basis to develop the same amount of oil and gas reserves as were developed in 1985.

Had technology remained constant since 1985, it would take two wells to produce the same amount of oil and natural gas as one 1985 well. However, advances in technology mean that one well today can produce two times as much as a single 1985 well.

Drilling wastes have decreased by as much as 148 million barrels due to increased well productivity and fewer wells.

The drilling footprint of well pads has decreased by as much as 70 percent due to advanced drilling technology, which is extremely useful for drilling in sensitive areas.

By using modular drilling rigs and slimhole drilling, the size and weight of drilling rigs can be reduced by up to 75 percent over traditional drilling rigs, reducing their surface impact.

Had technology, and thus drilling footprints, remained at 1985 levels, today's drilling footprints would take up an additional 17,000 acres of land.

New exploration techniques and vibrational sources mean less reliance on explosives, reducing the impact of exploration on the environment.

Some of the major recent technological innovations in the exploration and production sector include:

Advanced 3-D Seismic Imaging and 4-D Seismic Imaging

The development of seismic imaging in three dimensions greatly changed the nature of natural gas exploration. This technology uses traditional seismic imaging techniques, combined with powerful computers and processors, to create a three-dimensional model of the subsurface layers. 4-D seismology expands on this, by adding time as a dimension, allowing exploration teams to observe how subsurface characteristics change over time. Exploration teams can now identify natural gas prospects more easily, place wells more effectively, reduce the number of dry holes drilled, reduce drilling costs, and cut exploration time. This leads to both economic and environmental benefits.

CO2-Sand Fracturing—Fracturing techniques have been used since the 1970s to help increase the flow rate of natural gas and oil from underground formations. CO2-Sand fracturing involves using a mixture of sand proppants and liquid CO2 to fracture formations, creating and enlarging cracks through which oil and natural gas may flow more freely. The CO2 then vaporizes, leaving only sand in the formation, holding the newly enlarged cracks open. Because there are no other substances used in this type of fracturing, there are no 'leftovers' from the fracturing process that must be removed. This means that, while this type of fracturing effectively opens the formation and allows for increased recovery of oil and natural gas, it does not damage the deposit, generates no below ground wastes, and protects groundwater resources.

Coiled Tubing

Coiled tubing technologies replace the traditional rigid, jointed drill pipe with a long, flexible coiled pipe string. This greatly reduces the cost of drilling, as well as providing a smaller drilling footprint, requiring less drilling mud, faster rig set up, and reducing the time normally needed to make drill pipe connections. Coiled tubing can also be used in combination with slimhole drilling to provide very economic drilling conditions, and less impact on the environment.

Measurement While Drilling

Measurement-While-Drilling (MWD) systems allow for the collection of data from the bottom of a well as it is being drilled. This allows engineers and drilling teams access to up-to-the-second information on the exact nature of the rock formations being encountered by the drill bit. This improves drilling efficiency and accuracy in the drilling process, allows better formation evaluation as the drill bit encounters the underground formation, and reduces the chance of formation damage and blowouts.

Slimhole Drilling

Slimhole drilling is exactly as it sounds; drilling a slimmer hole in the ground to get to natural gas and oil deposits. In order to be considered slimhole drilling, at least 90 percent of a well must be drilled with a drill bit less than six inches in diameter (whereas conventional wells typically use drill bits as large as 12.25 inches in diameter). Slimhole drilling can significantly improve the efficiency of drilling operations, as well as decrease its environmental impact. In fact, shorter drilling times and smaller drilling crews can translate into a 50 percent reduction in drilling costs, while reducing the drilling footprint by as much as 75 percent. Because of its low cost profile and reduced environmental impact, slimhole drilling provides a method of economically drilling exploratory wells in new areas, drilling deeper wells in existing fields, and providing an efficient means for extracting more natural gas and oil from un-depleted fields.

Offshore Production—NASA of the Sea

The offshore oil and natural gas production sector is sometimes compared to the aeronautics field and NASA due to achievements in deepwater drilling that have been facilitated by state of the art technology. New technology, including improved offshore drilling rigs, dynamic positioning devices and sophisticated navigation systems are allowing safe, efficient offshore drilling in waters more than 10,000 feet deep. Visit the offshore drilling section to learn more.

Hydraulic Fracturing also called "Fracking," or "Frac'ing"—Used to free natural gas that is trapped in shale rock formations. A liquid mix that is 99 percent water and sand is injected into the rock at very high pressure, creating fractures within the rock that provide the natural gas a path to flow to the wellhead. The fracking fluid mix also helps to keep the formation more porous. Hydraulic fracturing is now widely used, with more than 90 percent of the natural gas wells in the United States having used it to boost production at some time.

The above technological advancements provide only a snapshot of the increasingly sophisticated technology being developed and put into practice in the exploration and production of natural gas and oil. New technologies and applications are being developed constantly, and serve to improve the economics of producing natural gas, allow for the production of deposits formerly considered too unconventional or uneconomic to develop, and ensure that the supply of natural gas keeps up with steadily increasing demand. Sufficient domestic natural gas resources exist to help fuel the U.S. for a significant period of time, and technology is playing a huge role in providing low-cost, environmentally sound methods of extracting these resources.

Two other technologies that are revolutionizing the natural gas industry include the increased use of liquefied natural gas, and natural gas fuel cells. These technologies are discussed below.

Liquefied Natural Gas

Cooling natural gas to about -260°F at normal pressure results in the condensation of the gas into liquid form, known as Liquefied Natural Gas (LNG). LNG can be very useful, particularly for the transportation of natural gas, since LNG takes up about one six hundredth the volume of gaseous natural gas. Advances in technology are reducing the costs associated with the liquification and regasification of LNG. Because it is easy to transport, LNG can serve to make economical stranded natural gas deposits from around the globe for which the construction of pipelines is uneconomical.

LNG Delivery Facility with Tanker

LNG, when vaporized to gaseous form, will only burn in concentrations of between 5 and 15 percent mixed with air. In addition, LNG, or any vapor associated with LNG, will not explode in an unconfined environment. Thus, in the unlikely event of an LNG spill, the natural gas has little chance of igniting an explosion. Liquification removes oxygen, carbon dioxide, sulfur, and water from the natural gas, resulting in LNG that is almost pure methane.

LNG is typically transported by specialized tanker with insulated walls, and is kept in liquid form by autorefrigeration, a process in which the LNG is kept at its boiling point, so that any heat additions are countered by the energy lost from LNG vapor that is vented out of storage and used to power the vessel.

The increased use of LNG is allowing for the production and marketing of natural gas deposits that were previously economically unrecoverable. Although it currently accounts for only about 1 percent of natural gas used in the United States, it is expected that LNG imports will provide a steady, dependable source of natural gas for U.S. consumption.

Natural Gas Fuel Cells

Fuel cells powered by natural gas are an extremely exciting and promising new technology for the clean and efficient generation of electricity. Fuel cells have the ability to generate electricity using electrochemical reactions as opposed to combustion of fossil fuels to generate electricity. Essentially, a fuel cell works by passing streams of fuel (usually hydrogen) and oxidants over electrodes that are separated by an electrolyte. This produces a chemical reaction that generates electricity without requiring the combustion of fuel, or the addition of heat as is common in the traditional generation of electricity. When pure hydrogen is used as fuel, and pure oxygen is used as the oxidant, the reaction that takes place within a fuel cell produces only water, heat, and electricity. In practice, fuel cells result in very low emission of harmful pollutants, and the generation of high-quality, reliable electricity. The use of natural gas-powered fuel cells has a number of benefits, including:

Clean Electricity

Fuel cells provide the cleanest method of producing electricity from fossil fuels. While a pure hydrogen, pure oxygen fuel cell produces only water, electricity, and heat, fuel cells in practice emit trace amounts of sulfur compounds and very low levels of carbon dioxide. However, the carbon dioxide produced by fuel cell use is concentrated and can be readily recaptured, as opposed to being emitted into the atmosphere.

Distributed Generation—Fuel cells can come in extremely compact sizes, allowing for their placement wherever electricity is needed. This includes residential, commercial, industrial, and even transportation settings.

Dependability—Fuel cells are completely enclosed units, with no moving parts or complicated machinery. This translates into a dependable source of electricity, capable of operating for thousands of hours. In addition, they are very quiet and safe sources of electricity. Fuel cells also do not have electricity surges, meaning they can be used where a constant, dependable source of electricity is needed.

Efficiency—Fuel cells convert the energy stored within fossil fuels into electricity much more efficiently than traditional generation of electricity using combustion. This means that less fuel is required to produce the same amount of electricity. The National Energy Technology Laboratory estimates that, used in combination with natural gas turbines, fuel cell generation facilities can be produced that will operate in the 1 to 20 Megawatt range at 70 percent efficiency, which is much higher than the efficiencies that can be reached by traditional generation methods within that output range.

How a Fuel Cell Works

The generation of electricity has traditionally been a very polluting, inefficient process. However, with new fuel cell technology, the future of electricity generation is expected to change dramatically in the next ten to twenty years. Research and development into fuel cell technology is ongoing, to ensure that the technology is refined to a level where it is cost-effective for all varieties of electric generation requirements.

Natural Gas Technology Resources

The natural gas industry is joined by government agencies and laboratories, private research and development firms, and environmental technology groups in coming up with new technologies that may improve the efficiency, cost-effectiveness, and environmental soundness of the natural gas industry. New technology and methods emerge frequently in the natural gas industry. Below are links to a number of resources that provide information on new technological developments in the oil and natural gas industry:

Over the past few decades the oil and natural gas industry has transformed into one of the most technologically advanced industries in the United States. New innovations have reshaped the industry into a technology leader. This section will discuss the role of technology in the evolution of the natural gas industry, focusing on technologies in the exploration and production sector, as well as a few select innovations that have had a profound effect on the potential for natural gas. Scroll down, or click on the links below to jump ahead:Beginning in the 1990s, more power generators began to use natural gas to make electricity, causing demand for natural gas to grow substantially. The natural gas industry has been able to keep pace with growing demand and produce greater amounts of natural gas through technological innovation. These innovations have enabled the development of natural gas from shale and other formerly "unconventional" formations that are found in abundance across the United States, as well as development from traditional offshore and onshore formations. Below is a brief list of some of the major technological advancements that have been made recently:

Beginning in the 1990s, more power generators began to use natural gas to make electricity, causing demand for natural gas to grow substantially. The natural gas industry has been able to keep pace with growing demand and produce greater amounts of natural gas through technological innovation. These innovations have enabled the development of natural gas from shale and other formerly "unconventional" formations that are found in abundance across the United States, as well as development from traditional offshore and onshore formations. Below is a brief list of some of the major technological advancements that have been made recently:

According to a Department of Energy Report, "Environmental Benefits of Advanced Oil and Gas Exploration and Production Technology," released in 1999 and still one of the most indepth analyses available as of 2010:

According to a Department of Energy Report, "Environmental Benefits of Advanced Oil and Gas Exploration and Production Technology," released in 1999 and still one of the most indepth analyses available as of 2010:

22,000 fewer wells are needed on an annual basis to develop the same amount of oil and gas reserves as were developed in 1985. In addition, it would take two wells to produce the same amount of oil and natural gas as one 1985 well. However, advances in technology mean that one well today can produce two times as much as a single 1985 well. The drilling footprint of well pads has decreased by as much as 70 percent due to advanced drilling technology, which is extremely useful for drilling in sensitive areas.

Had technology, and thus drilling footprints, remained at 1985 levels, today's drilling footprints would take up an additional 17,000 acres of land.

New exploration techniques and vibrational sources mean less reliance on explosives, reducing the impact of exploration on the environment.

Transportation Process and Flow

Overview | Gathering System | Processing Plant | Transmission Grid | Market Centers/Hubs | Underground Storage | Peak Shaving

Overview

Transporting natural gas from the wellhead to the final customer involves several physical transfers of custody and multiple processing steps. A natural gas pipeline system begins at the natural gas producing well or field. Once the gas leaves the producing well, a pipeline gathering system directs the flow either to a natural gas processing plant or directly to the mainline transmission grid, depending upon the initial quality of the wellhead product.

The processing plant produces pipeline-quality natural gas. This gas is then transported by pipeline to consumers or is put into underground storage for future use. Storage helps to maintain pipeline system operational integrity and/or to meet customer requirements during peak-usage periods.

Transporting natural gas from wellhead to market involves a series of processes and an array of physical facilities. Among these are:

Gathering Lines—These small-diameter pipelines move natural gas from the wellhead to the natural gas processing plant or to an interconnection with a larger mainline pipeline.

Processing Plant—This operation extracts natural gas liquids and impurities from the natural gas stream.

Mainline Ttransmission Systems—These wide-diameter, long-distance pipelines transport natural gas from the producing area to market areas.

Market Hubs/Centers—Locations where pipelines intersect and flows are transferred.

Underground Storage Facilities—Natural gas is stored in depleted oil and gas reservoirs, aquifers, and salt caverns for future use.

Peak Shaving—System design methodology permitting a natural gas pipeline to meet short-term surges in customer demands with minimal infrastructure. Peaks can be handled by using gas from storage or by short-term line-packing.

The Natural Gas Gathering System

A natural gas pipeline system begins at a natural gas producing well or field. In the producing area many of the pipeline systems are primarily involved in "gathering" operations. That is, a pipeline is connected to a producing well, converging with pipes from other wells where the natural gas stream may be subjected to an extraction process to remove water and other impurities if needed. Natural gas exiting the production field is usually referred to as "wet" natural gas if it still contain significant amounts of hydrocarbon liquids and contaminants.

Under certain conditions some or all of the natural gas produced at a well may be returned to the reservoir in cycling, repressuring, or conservation operations and/or vented and flared. At this stage it is a mixture of methane and other hydrocarbons, as well as some non-hydrocarbons, existing in the gaseous phase or in a solution with crude oil. The principal hydrocarbons normally contained in the natural gas mixture are methane, ethane, propane, butane, and pentane. Typical non-hydrocarbon gases that may be present in reservoir natural gas are water vapor, carbon dioxide, helium, hydrogen sulfide, and nitrogen.

Gathering Lines—These small-diameter pipelines move natural gas from the wellhead to the natural gas processing plant or to an interconnection with a larger mainline pipeline.

Processing Plant—This operation extracts natural gas liquids and impurities from the natural gas stream.

Mainline Ttransmission Systems—These wide-diameter, long-distance pipelines transport natural gas from the producing area to market areas.

Market Hubs/Centers—Locations where pipelines intersect and flows are transferred.

Underground Storage Facilities—Natural gas is stored in depleted oil and gas reservoirs, aquifers, and salt caverns for future use.

Peak Shaving—System design methodology permitting a natural gas pipeline to meet short-term surges in customer demands with minimal infrastructure. Peaks can be handled by using gas from storage or by short-term line-packing.

Gathering Lines—These small-diameter pipelines move natural gas from the wellhead to the natural gas processing plant or to an interconnection with a larger mainline pipeline.

Processing Plant—This operation extracts natural gas liquids and impurities from the natural gas stream.

Mainline Ttransmission Systems—These wide-diameter, long-distance pipelines transport natural gas from the producing area to market areas.

Market Hubs/Centers—Locations where pipelines intersect and flows are transferred.

Underground Storage Facilities—Natural gas is stored in depleted oil and gas reservoirs, aquifers, and salt caverns for future use.

Peak Shaving—System design methodology permitting a natural gas pipeline to meet short-term surges in customer demands with minimal infrastructure. Peaks can be handled by using gas from storage or by short-term line-packing.

Gathering Lines—These small-diameter pipelines move natural gas from the wellhead to the natural gas processing plant or to an interconnection with a larger mainline pipeline.

Processing Plant—This operation extracts natural gas liquids and impurities from the natural gas stream.

Mainline Ttransmission Systems—These wide-diameter, long-distance pipelines transport natural gas from the producing area to market areas.

Market Hubs/Centers—Locations where pipelines intersect and flows are transferred.

Underground Storage Facilities—Natural gas is stored in depleted oil and gas reservoirs, aquifers, and salt caverns for future use.

Peak Shaving—System design methodology permitting a natural gas pipeline to meet short-term surges in customer demands with minimal infrastructure. Peaks can be handled by using gas from storage or by short-term line-packing.

Gathering Lines—These small-diameter pipelines move natural gas from the wellhead to the natural gas processing plant or to an interconnection with a larger mainline pipeline.

Processing Plant—This operation extracts natural gas liquids and impurities from the natural gas stream.

Mainline Ttransmission Systems—These wide-diameter, long-distance pipelines transport natural gas from the producing area to market areas.

Market Hubs/Centers—Locations where pipelines intersect and flows are transferred.

Underground Storage Facilities—Natural gas is stored in depleted oil and gas reservoirs, aquifers, and salt caverns for future use.

Peak Shaving—System design methodology permitting a natural gas pipeline to meet short-term surges in customer demands with minimal infrastructure. Peaks can be handled by using gas from storage or by short-term line-packing.

Gathering Lines—These small-diameter pipelines move natural gas from the wellhead to the natural gas processing plant or to an interconnection with a larger mainline pipeline.

Processing Plant—This operation extracts natural gas liquids and impurities from the natural gas stream.

Mainline Ttransmission Systems—These wide-diameter, long-distance pipelines transport natural gas from the producing area to market areas.

Market Hubs/Centers—Locations where pipelines intersect and flows are transferred.

Underground Storage Facilities—Natural gas is stored in depleted oil and gas reservoirs, aquifers, and salt caverns for future use.

Peak Shaving—System design methodology permitting a natural gas pipeline to meet short-term surges in customer demands with minimal infrastructure. Peaks can be handled by using gas from storage or by short-term line-packing.

Gathering Lines—These small-diameter pipelines move natural gas from the wellhead to the natural gas processing plant or to an interconnection with a larger mainline pipeline.

Processing Plant—This operation extracts natural gas liquids and impurities from the natural gas stream.

Mainline Ttransmission Systems—These wide-diameter, long-distance pipelines transport natural gas from the producing area to market areas.

Market Hubs/Centers—Locations where pipelines intersect and flows are transferred.

Underground Storage Facilities—Natural gas is stored in depleted oil and gas reservoirs, aquifers, and salt caverns for future use.

Peak Shaving—System design methodology permitting a natural gas pipeline to meet short-term surges in customer demands with minimal infrastructure. Peaks can be handled by using gas from storage or by short-term line-packing.

The processing plant produces pipeline-quality natural gas. This gas is then transported by pipeline to consumers or is put into underground storage for future use. Storage helps to maintain pipeline system operational integrity and/or to meet customer requirements during peak-usage periods.

Transporting natural gas from wellhead to market involves a series of processes and an array of physical facilities. Among these are:

Gathering Lines—These small-diameter pipelines move natural gas from the wellhead to the natural gas processing plant or to an interconnection with a larger mainline pipeline.

Processing Plant—This operation extracts natural gas liquids and impurities from the natural gas stream.

Mainline Ttransmission Systems—These wide-diameter, long-distance pipelines transport natural gas from the producing area to market areas.

Market Hubs/Centers—Locations where pipelines intersect and flows are transferred.

Underground Storage Facilities—Natural gas is stored in depleted oil and gas reservoirs, aquifers, and salt caverns for future use.

Peak Shaving—System design methodology permitting a natural gas pipeline to meet short-term surges in customer demands with minimal infrastructure. Peaks can be handled by using gas from storage or by short-term line-packing.

The Natural Gas Gathering System

A natural gas pipeline system begins at a natural gas producing well or field. In the producing area many of the pipeline systems are primarily involved in "gathering" operations. That is, a pipeline is connected to a producing well, converging with pipes from other wells where the natural gas stream may be subjected to an extraction process to remove water and other impurities if needed. Natural gas exiting the production field is usually referred to as "wet" natural gas if it still contain significant amounts of hydrocarbon liquids and contaminants.

Under certain conditions some or all of the natural gas produced at a well may be returned to the reservoir in cycling, repressuring, or conservation operations and/or vented and flared. At this stage it is a mixture of methane and other hydrocarbons, as well as some non-hydrocarbons, existing in the gaseous phase or in a solution with crude oil. The principal hydrocarbons normally contained in the natural gas mixture are methane, ethane, propane, butane, and pentane. Typical non-hydrocarbon gases that may be present in reservoir natural gas are water vapor, carbon dioxide, helium, hydrogen sulfide, and nitrogen.

In proximity to the well are facilities that produce what is referred to as "lease condensate", that is, a mixture consisting primarily of pentanes and heavier hydrocarbons which is recovered as a liquid from natural gas. Other natural gas liquids, such as butane and propane, are recovered at downstream natural gas processing plants or facilities (see below).

The principal service provided by a natural gas processing plant to the natural gas mainline transmission network is that it produces pipeline quality natural gas. Natural gas mainline transmission systems are designed to operate within certain tolerances. Natural gas entering the system that is not within certain specific gravities, pressures, Btu content range, or water content level will cause operational problems, pipeline deterioration, or even cause pipeline rupture.

Natural gas processing plants are also facilities designed to recover natural gas liquids from a stream of natural gas that may or may not have passed through lease separators and/or field separation facilities. These facilities also control the quality of the natural gas to be marketed. Several types of natural gas processing plants, employing various techniques and technologies to extract contaminants and natural gas liquids, are used to produce pipeline quality "dry" gas. At many processing plants the primary objective is the production of dry gas (demethanizing). Any remaining natural gas liquids extraction stream is directed to a separate plant to undergo what is referred to as a "fractionation" process.

But a number of natural gas processing plants do include these fractionation facilities, where saturated hydrocarbons are removed from natural gas and separated into distinct parts, or "fractions," such as propane, butane, and ethane. Essentially, natural gas is methane, a colorless, odorless, flammable hydrocarbon gas (CH_4). Also present in natural gas production, especially that in association with oil production, are a number of petroleum gases. They include (in addition to ethane, propane and butane) ethylene, propylene, butylene, isobutane, and isobutylene. They are derived from crude oil refining or natural gas fractionation and are liquefied through pressurization.

A natural gas mainline system will tend to be designed as either a grid or a trunkline system. The latter is usually a long-distance, wide-diameter pipeline system that generally links a major supply source with a market area or with a large pipeline/LDC serving a market area. Trunklines tend to have fewer receipt points (usually at the beginning of its route), fewer delivery points, interconnections with other pipelines, and associated lateral lines.

Between the producing area, or supply source, and the market area, a number of compressor stations are located along the transmission system. These stations contain one or more compressor units whose purpose is to receive the transmission flow (which has decreased in pressure since the previous compressor station) at an intake point, increase the pressure and rate of flow, and thus, maintain the movement of natural gas along the pipeline.

Between the producing area, or supply source, and the market area, a number of compressor stations are located along the transmission system. These stations contain one or more compressor units whose purpose is to receive the transmission flow (which has decreased in pressure since the previous compressor station) at an intake point, increase the pressure and rate of flow, and thus, maintain the movement of natural gas along the pipeline.

Between the producing area, or supply source, and the market area, a number of compressor stations are located along the transmission system. These stations contain one or more compressor units whose purpose is to receive the transmission flow (which has decreased in pressure since the previous compressor station) at an intake point, increase the pressure and rate of flow, and thus, maintain the movement of natural gas along the pipeline.

Between the producing area, or supply source, and the market area, a number of compressor stations are located along the transmission system. These stations contain one or more compressor units whose purpose is to receive the transmission flow (which has decreased in pressure since the previous compressor station) at an intake point, increase the pressure and rate of flow, and thus, maintain the movement of natural gas along the pipeline.

To address the potential for pipeline rupture, safety cutoff meters are installed along a mainline transmission system route. Devices located at strategic points are designed to detect a drop in pressure that would result from a downstream or upstream pipeline rupture and automatically stop the flow of natural gas beyond its location. Monitoring the pipeline as a whole are apparatus known as (SCADA Systems Control and Data Acquisition) systems. SCADA systems provide monitoring staff the ability to direct and control pipeline flows, maintaining pipeline integrity and pressures as natural gas is received and delivered along numerous points on the system, including flows into and out of storage facilities.

To address the potential for pipeline rupture, safety cutoff meters are installed along a mainline transmission system route. Devices located at strategic points are designed to detect a drop in pressure that would result from a downstream or upstream pipeline rupture and automatically stop the flow of natural gas beyond its location. Monitoring the pipeline as a whole are apparatus known as (SCADA Systems Control and Data Acquisition) systems. SCADA systems provide monitoring staff the ability to direct and control pipeline flows, maintaining pipeline integrity and pressures as natural gas is received and delivered along numerous points on the system, including flows into and out of storage facilities.

Richard L. Itteilag

Home > Natural Gas > About U.S. Natural Gas Pipelines > Regional/State Underground Natural Gas Storage Table

About U.S. Natural Gas Pipelines—Transporting Natural Gas based on data through 2007/2008 with selected updates

Regional Underground Natural Gas Storage, Close of 2007

	Depleted-Reservoir Storage	Aquifer Storage	Salt-Cavern Storage	Total Region/ State	# of Sites	Working Gas Capacity (Bcf)	Daily Withdrawal Capability (MMcf)	# of Sites	Working Gas Capacity (Bcf)	Daily Withdrawal Capability (MMcf)	# of Sites	Working Gas Capacity (Bcf)	Daily Withdrawal Capability (MMcf)
Central Region													
Colorado	8	42	1,088	0	0	0	0	0	0	0	8	42	1,088
Iowa	0	0	0	4	77	1,060	0	0	0	4	77	1,060	
Kansas	18	116	2,418	0	0	0	1	1	0	19	117	2,418	
Missouri	0	0	0	1	11	350	0	0	0	1	11	350	
Montana	5	196	310	0	0	0	0	0	0	5	196	310	
Nebraska	1	16	169	0	0	0	0	0	0	1	16	169	
Utah	1	51	427	2	1	100	0	0	0	3	52	527	
Wyoming	7	45	227	1	1	75	0	0	0	8	46	302	
Total Sites	40	466	4,639	8	90	1,585	1	1	0	49	557	6,224	
(Marginal Sites)1	(5)	(2)	(186)	(0)	(0)	(0)	(1)	(1)	(0)	(6)	(3)	(186)	
Midwest Region													
Illinois	11	52	833	18	249	5,294	0	0	0	29	301	6,136	
Indiana	10	14	261	12	20	526	0	0	0	22	34	787	
Michigan	43	641	16,786	0	0	0	2	2	85	45	643	16,871	
Minnesota	0	0	0	1	2	60	0	0	0	1	2	60	
Ohio	24	220	4,670	0	0	0	0	0	0	24	220	4,670	
Total Sites	88	927	22,549	31	271	5,855	2	2	85	121	1,200	28,524	
(Marginal Sites)1	(10)	(8)	(258)	(2)	(4)	(108)	(0)	(0)	(0)	(12)	(12)	(366)	
Northeast Region													
Maryland	1	17	400	0	0	0	0	0	0	1	17	400	
New York	23	116	1,696	0	0	0	1	1	145	24	117	1,841	
Pennsylvania	50	406	8,615	0	0	0	0	0	0	50	406	8,615	

Virginia	1	1	20	0	0	0	2	4	325	3	5	345
West Virginia	32	251	4,002	0	0	0	0	0	0	32	251	4,002
Total Sites	107	791	14,733	0	0	0	3	5	470	110	796	15,203
(Marginal Sites)[1]	(7)	(31)	(43)	(0)	(0)	(0)	(0)	(0)	(0)	(7)	(31)	(43)
Southeast Region												
Alabama	1	8	300	0	0	0	1	7	600	2	15	900
Kentucky	20	89	1,792	3	7	68	0	0	0	23	96	1,860
Mississippi	5	51	1,048	0	0	0	3	32	3,022	8	83	4,070
Tennessee	1	1	20	0	0	0	0	0	0	1	1	20
Total Sites	27	149	3,160	3	7	68	4	39	4,115	34	195	6,850
(Marginal Sites)[1]	(4)	(2)	(24)	(0)	(0)	(0)	(0)	(0)	(0)	(4)	(2)	(24)
Southwest Region												
Arkansas	2	15	231	0	0	0	0	0	0	2	15	231
Louisiana	8	286	3,965	0	0	0	7	48	3,165	15	334	7,130
New Mexico	2	54	375	0	0	0	0	0	0	2	54	375
Oklahoma	13	194	3,772	0	0	0	0	0	0	13	194	3,772
Texas	20	365	5,334	0	0	0	14	78	6,,361	34	443	11,695
Total Sites	45	914	13,677	0	0	0	21	126	9,526	66	1,040	23,203
(Marginal Sites)[1]	(8)	(30)	(184)	(0)	(0)	(0)	(1)	(4)	(0)	(9)	(34)	(184)
Western Region												
California	12	266	6,875	0	0	0	0	0	0	12	266	6,875
Oregon	7	15	497	0	0	0	0	0	0	7	15	497
Washington	0	0	0	1	22	850	0	0	0	1	22	850
Total Sites	19	281	7,372	1	22	850	0	0	0	20	303	8,222
(Marginal Sites)[1]	(0)	(0)	(0)	(0)	(0)	(0)	(0)	(0)	(0)	(0)	(0)	(0)
Total U.S. Sites	326	3,528	66,130	43	390	8,393	31	173	13,703	400	4,091	88,226
(Marginal Sites)[1]	(34)	(73)	(695)	(2)	(4)	(108)	(2)	(5)	(0)	(38)	(82)	(803)

[1] Marginal sites: very little or no activity reported during the 2007 calendar year. However, these sites are included in regional and State summary totals.

Note: Bcf = Billion cubic feet. MMcf = Million cubic feet. Totals may not sum due to independent rounding.

Source: Energy Information Administration, GasTran Natural Gas Transportation Information System, Underground Natural Gas Storage Database, December 2008.

http://205.254.135.7/pub/oil_gas/natural_gas/analysis_publications/ngpipeline/UGTable.htmlhttp://205.254.135.7/pub/oil_gas/analysis_publications/ngpipeline/UGTable.htmlhttp://205.254.135.7/pub/oil_gas/natural_gas/analysis_publications/ngpipeline/UGTable.htmlhttp://205.254.135.7/pub/oil_gas/natural_gas/analysis_publications/ngpipeline/UGTable.htmlMost underground storage facilities, 327 out of 399 at the beginning of 2008, are depleted reservoirs, which are close to consumption centers and which were relatively easy to convert to storage service. In some areas, however, most notably the Midwestern United States, some natural aquifers have been converted to natural gas storage reservoirs. An aquifer is suitable for natural gas storage if the water-bearing sedimentary rock formation is overlaid with an impermeable cap rock. While the geology of aquifers is similar to that of depleted production fields, their use in natural gas storage usually requires more base (cushion) gas and greater monitoring of withdrawal and injection performance. Deliverability rates may be enhanced by the presence of an active water drive.

But a number of natural gas processing plants do include these fractionation facilities, where saturated hydrocarbons are removed from natural gas and separated into distinct parts, or "fractions," such as propane, butane, and ethane. Essentially, natural gas is methane, a colorless, odorless, flammable hydrocarbon gas (CH4). Also present in natural gas production, especially that in association with oil production, are a number of petroleum gases. They include (in addition to ethane, propane and butane) ethylene, propylene, butylene, isobutane, and isobutylene. They are derived from crude oil refining or natural gas fractionation and are liquefied through pressurization.

But a number of natural gas processing plants do include these fractionation facilities, where saturated hydrocarbons are removed from natural gas and separated into distinct parts, or "fractions," such as propane, butane, and ethane. Essentially, natural gas is methane, a colorless, odorless, flammable hydrocarbon gas (CH4). Also present in natural gas production, especially that in association with oil production, are a number of petroleum gases. They include (in addition to ethane, propane and butane) ethylene, propylene, butylene, isobutane, and isobutylene. They are derived from crude oil refining or natural gas fractionation and are liquefied through pressurization.

In proximity to the well are facilities that produce what is referred to as "lease condensate", that is, a mixture consisting primarily of pentanes and heavier hydrocarbons which is recovered as a liquid from natural gas. Other natural gas liquids, such as butane and propane, are recovered at downstream natural gas processing plants or facilities (see below).

Once it leaves the producing area, a pipeline system directs flow either to a natural gas processing plant or directly to the mainline transmission grid. Nonassociated natural gas, that is, natural gas that is not in contact with significant quantities of crude oil in the reservoir, is sometimes of pipeline quality after undergoing a decontamination process in the production area, and does not need to flow through a processing plant prior to entering the mainline transmission system.

The Natural Gas Processing Plant

The principal service provided by a natural gas processing plant to the natural gas mainline transmission network is that it produces pipeline quality natural gas. Natural gas mainline transmission systems are designed to operate within certain tolerances. Natural gas entering the system that is not within certain specific gravities, pressures, Btu content range, or water content level will cause operational problems, pipeline deterioration, or even cause pipeline rupture.

Natural gas processing plants are also facilities designed to recover natural gas liquids from a stream of natural gas that may or may not have passed through lease separators and/or field separation facilities. These facilities also control the quality of the natural gas to be marketed. Several types of natural gas processing plants, employing various techniques and technologies to extract contaminants and natural gas liquids, are used to produce pipeline quality "dry" gas. At many processing plants the primary objective is the production of dry gas (demethanizing). Any remaining natural gas liquids extraction stream is directed to a separate plant to undergo what is referred to as a "fractionation" process.

But a number of natural gas processing plants do include these fractionation facilities, where saturated hydrocarbons are removed from natural gas and separated into distinct parts, or "fractions," such as propane, butane, and ethane. Essentially, natural gas is methane, a colorless, odorless, flammable hydrocarbon gas (CH_4). Also present in natural gas production, especially that in association with oil production, are a number of petroleum gases. They include (in addition to ethane, propane and butane) ethylene, propylene, butylene, isobutane, and isobutylene. They are derived from crude oil refining or natural gas fractionation and are liquefied through pressurization.

The Transmission Grid and Compressor Stations

The natural gas mainline (transmission line) is a wide-diameter, often-times long-distance, portion of a natural gas pipeline system, excluding laterals, located between the gathering system (production area), natural gas processing plant, other receipt points, and the principal customer service area(s). The lateral, usually of smaller diameter, branches off the mainline natural gas pipeline to connect with or serve a specific customer or group of customers.

A natural gas mainline system will tend to be designed as either a grid or a trunkline system. The latter is usually a long-distance, wide-diameter pipeline system that generally links a major supply source with a market area or with a large pipeline/LDC serving a market area. Trunklines tend to have fewer receipt points (usually at the beginning of its route), fewer delivery points, interconnections with other pipelines, and associated lateral lines.

A grid type transmission system is usually characterized by a large number of laterals or branches from the mainline, which tend to form a network of integrated receipt, delivery and pipeline interconnections that operate in, and serve major market areas. In form, they are similar to a local distribution company (LDC) network configuration, but on a much larger scale.

Between the producing area, or supply source, and the market area, a number of compressor stations are located along the transmission system. These stations contain one or more compressor units whose purpose is to receive the transmission flow (which has decreased in pressure since the previous compressor station) at an intake point, increase the pressure and rate of flow, and thus, maintain the movement of natural gas along the pipeline.

Compressor units that are used on a natural gas mainline transmission system are usually rated at 1,000 horsepower or more and are of the centrifugal (turbine) or reciprocating (piston) type. The larger compressor stations may have as many as 10-16 units with an overall horsepower rating of from 50,000 to 80,000 HP and a throughput capacity exceeding three billion cubic feet of natural gas per day. Most compressor units operate on natural gas (extracted from the pipeline flow); but in recent years, and mainly for environmental reasons, the use of electricity driven compressor units has been growing.

Many of the larger mainline transmission routes are what is generally referred to as "looped." Looping is when one pipeline is laid parallel to another and is often used as a way to increase capacity along a right-of-way beyond what is possible on one line, or an expansion of an existing pipeline(s). These lines are connected to move a larger flow along a single segment of the pipeline system. Some very large pipeline systems have 5 or 6 large diameter pipes laid along the same right-of-way. Looped pipes may extend the distance between compressor stations, where they can transfer part of their flow, or the looping may be limited to only a portion of the line between stations. In the latter case, the looping often serves as essentially a storage device, where natural gas can be line-packed as a way to increase deliveries to local customers during certain peak periods.

To address the potential for pipeline rupture, safety cutoff meters are installed along a mainline transmission system route. Devices located at strategic points are designed to detect a drop in pressure that would result from a downstream or upstream pipeline rupture and automatically stop the flow of natural gas beyond its location. Monitoring the pipeline as a whole are apparatus known as (SCADA Systems Control and Data Acquisition) systems. SCADA systems provide monitoring staff the ability to direct and control pipeline flows, maintaining pipeline integrity and pressures as natural gas is received and delivered along numerous points on the system, including flows into and out of storage facilities.

Natural Gas Market Centers/Hubs

Natural gas market centers and hubs evolved, beginning in the late 1980s, as an outgrowth of natural gas market restructuring and the execution of a number of Federal Energy Regulatory Commission's (FERC) Orders culminating in Order 636 issued in 1992. Order 636 mandated that interstate natural gas pipeline companies transform themselves from buyers and sellers of natural gas to strictly natural gas transporters. Market centers and hubs were developed to provide new natural gas shippers with many of the physical capabilities and administrative support services formally handled by the interstate pipeline company as "bundled" sales services.

Two key services offered by market centers/hubs are transportation between and interconnections with other pipelines and the physical coverage of short-term receipt/delivery balancing needs. Many of these centers also provide unique services that help expedite and improve the natural gas transportation process overall, such as Internet-based access to natural gas trading platforms and capacity release programs. Most also provide title transfer services between parties that buy, sell, or move their natural gas through the center.

As of the end of 2008, there were a total of 33 operational market centers in the United States (24) and Canada (9).

Underground Storage Facilities

At the end of the mainline transmission system, and sometimes at its beginning and in between, underground natural gas storage and LNG (liquefied natural gas) facilities provide for inventory management, supply backup, and the access to natural gas to maintain the balance of the system. There are three principal types of underground storage sites used in the United States today: depleted reservoirs in oil and/or gas fields, aquifers, and salt cavern formations. In one or two cases mine caverns have been used. Two of the most important characteristics of an underground storage reservoir are the capability to hold natural gas for future use, and the rate at which natural gas inventory can be injected and withdrawn (its deliverability rate).

Most underground storage facilities, 327 out of 399 at the beginning of 2008, are depleted reservoirs, which are close to consumption centers and which were relatively easy to convert to storage service. In some areas, however, most notably the Midwestern United States, some natural aquifers have been converted to natural gas storage reservoirs. An aquifer is suitable for natural gas storage if the water-bearing sedimentary rock formation is overlaid with an impermeable cap rock. While the geology of aquifers is similar to that of depleted production fields, their use in natural gas storage usually requires more base (cushion) gas and greater monitoring of withdrawal and injection performance. Deliverability rates may be enhanced by the presence of an active water drive.

During the past 20 years, the number of salt cavern storage sites has grown significantly because of its rapid cycling (inventory turnover) capability coupled with its ability to respond to daily, even hourly, variations in customer needs. The large majority of salt cavern storage facilities have been developed in salt dome formations located in the Gulf Coast States. Salt caverns leached from bedded salt formations in Northeastern, Midwestern, and Western States have also been developed but the number has been limited due to a lack of suitable geology. Cavern construction is more costly than depleted field conversions when measured on the basis of dollars per thousand cubic feet of working gas capacity, but the ability to perform several withdrawal and injection cycles each year reduces the per-unit cost of each thousand cubic feet of natural gas injected and withdrawn.

Peak Shaving

Underground natural gas storage inventories provide suppliers with the means to meet peak customer requirements up to a point. Beyond that point the distribution system still must be capable of meeting customer short-term peaking and volatile swing demands that occur on a daily and even hourly basis. During periods of extreme usage, peaking facilities, as well as other sources of temporary storage, are relied upon to supplement system and underground storage supplies.

Peaking needs are met in several ways. Some underground storage sites are designed to provide peaking service, but most often LNG (liquefied natural gas) in storage and liquefied petroleum gas such as propane are vaporized and injected into the natural gas distribution system supply to meet instant requirements. Short-term linepacking is also used to meet anticipated surge requirements.

The use of peaking facilities, as well as underground storage, is essentially a risk-management calculation, known as peak-shaving. The cost of installing these facilities is such that the incremental cost per unit is expensive. However, the cost of a service interruption, as well as the cost to an industrial customer in lost production, may be much higher. In the case of underground storage, a suitable site may not be locally available. The only other alternative might be to build or reserve the needed additional capacity on the pipeline network. Each alternative entails a cost.

A local natural gas distribution company (LDC) relies on supplemental supply sources (underground storage, LNG, and propane) and uses linepacking to "shave" as much of the difference between the total maximum user requirements (on a peak day or shorter period) and the baseload customer requirements (the normal or average) daily usage. Each unit "shaved" represents less demand charges (for reserving pipeline capacity on the trunklines between supply and market areas) that the LDC must pay. The objective is to maintain sufficient local underground natural gas storage capacity and have in place additional supply sources such as LNG and propane air to meet large shifts in daily demand, thereby minimizing capacity reservation costs on the supplying pipeline.

In proximity to the well are facilities that produce what is referred to as "lease condensate", that is, a mixture consisting primarily of pentanes and heavier hydrocarbons which is recovered as a liquid from natural gas. Other natural gas liquids, such as butane and propane, are recovered at downstream natural gas processing plants or facilities (see below).

Once it leaves the producing area, a pipeline system directs flow either to a natural gas processing plant or directly to the mainline transmission grid. Nonassociated natural gas, that is, natural gas that is not in contact with significant quantities of crude oil in the reservoir, is sometimes of pipeline quality after undergoing a decontamination process in the production area, and does not need to flow through a processing plant prior to entering the mainline transmission system.

The Natural Gas Processing Plant

The principal service provided by a natural gas processing plant to the natural gas mainline transmission network is that it produces pipeline quality natural gas. Natural gas mainline transmission systems are designed to operate within certain tolerances. Natural gas entering the system that is not within certain specific gravities, pressures, Btu content range, or water content level will cause operational problems, pipeline deterioration, or even cause pipeline rupture.

Natural gas processing plants are also facilities designed to recover natural gas liquids from a stream of natural gas that may or may not have passed through lease separators and/or field separation facilities. These facilities also control the quality of the natural gas to be marketed. Several types of natural gas processing plants, employing various techniques and technologies to extract contaminants and natural gas liquids, are used to produce pipeline quality "dry" gas. At many processing plants the primary objective is the production of dry gas (demethanizing). Any remaining natural gas liquids extraction stream is directed to a separate plant to undergo what is referred to as a "fractionation" process.

But a number of natural gas processing plants do include these fractionation facilities, where saturated hydrocarbons are removed from natural gas and separated into distinct parts, or "fractions," such as propane, butane, and ethane. Essentially, natural gas is methane, a colorless, odorless, flammable hydrocarbon gas (CH_4). Also present in natural gas production, especially that in association with oil production, are a number of petroleum gases. They include (in addition to ethane, propane and butane) ethylene, propylene, butylene, isobutane, and isobutylene. They are derived from crude oil refining or natural gas fractionation and are liquefied through pressurization.

The Transmission Grid and Compressor Stations

The natural gas mainline (transmission line) is a wide-diameter, often-times long-distance, portion of a natural gas pipeline system, excluding laterals, located between the gathering system (production area), natural gas processing plant, other receipt points, and the principal customer service area(s). The lateral, usually of smaller diameter, branches off the mainline natural gas pipeline to connect with or serve a specific customer or group of customers.

A natural gas mainline system will tend to be designed as either a grid or a trunkline system. The latter is usually a long-distance, wide-diameter pipeline system that generally links a major supply source with a market area or with a large pipeline/LDC serving a market area. Trunklines tend to have fewer receipt points (usually at the beginning of its route), fewer delivery points, interconnections with other pipelines, and associated lateral lines.

A grid type transmission system is usually characterized by a large number of laterals or branches from the mainline, which tend to form a network of integrated receipt, delivery and pipeline interconnections that operate in, and serve major market areas. In form, they are similar to a local distribution company (LDC) network configuration, but on a much larger scale.

Between the producing area, or supply source, and the market area, a number of compressor stations are located along the transmission system. These stations contain one or more compressor units whose purpose is to receive the transmission flow (which has decreased in pressure since the previous compressor station) at an intake point, increase the pressure and rate of flow, and thus, maintain the movement of natural gas along the pipeline.

Compressor units that are used on a natural gas mainline transmission system are usually rated at 1,000 horsepower or more and are of the centrifugal (turbine) or reciprocating (piston) type. The larger compressor stations may have as many as 10-16 units with an overall horsepower rating of from 50,000 to 80,000 HP and a throughput capacity exceeding three billion cubic feet of natural gas per day. Most compressor units operate on natural gas (extracted from the pipeline flow); but in recent years, and mainly for environmental reasons, the use of electricity driven compressor units has been growing.

Many of the larger mainline transmission routes are what is generally referred to as "looped." Looping is when one pipeline is laid parallel to another and is often used as a way to increase capacity along a right-of-way beyond what is possible on one line, or an expansion of an existing pipeline(s). These lines are connected to move a larger flow along a single segment of the pipeline system. Some very large pipeline systems have 5 or 6 large diameter pipes laid along the same right-of-way. Looped pipes may extend the distance between compressor stations, where they can transfer part of their flow, or the looping may be limited to only a portion of the line between stations. In the latter case, the looping often serves as essentially a storage device, where natural gas can be line-packed as a way to increase deliveries to local customers during certain peak periods.

To address the potential for pipeline rupture, safety cutoff meters are installed along a mainline transmission system route. Devices located at strategic points are designed to detect a drop in pressure that would result from a downstream or upstream pipeline rupture and automatically stop the flow of natural gas beyond its location. Monitoring the pipeline as a whole are apparatus known as (SCADA Systems Control and Data Acquisition) systems. SCADA systems provide monitoring staff the ability to direct and control pipeline flows, maintaining pipeline integrity and pressures as natural gas is received and delivered along numerous points on the system, including flows into and out of storage facilities.

Natural Gas Market Centers/Hubs

Natural gas market centers and hubs evolved, beginning in the late 1980s, as an outgrowth of natural gas market restructuring and the execution of a number of Federal Energy Regulatory Commission's (FERC) Orders culminating in Order 636 issued in 1992. Order 636 mandated that interstate natural gas pipeline companies transform themselves from buyers and sellers of natural gas to strictly natural gas transporters. Market centers and hubs were developed to provide new natural gas shippers with many of the physical capabilities and administrative support services formally handled by the interstate pipeline company as "bundled" sales services.

Two key services offered by market centers/hubs are transportation between and interconnections with other pipelines and the physical coverage of short-term receipt/delivery balancing needs. Many of these centers also provide unique services that help expedite and improve the natural gas transportation process overall, such as Internet-based access to natural gas trading platforms and capacity release programs. Most also provide title transfer services between parties that buy, sell, or move their natural gas through the center.

As of the end of 2008, there were a total of 33 operational market centers in the United States (24) and Canada (9).

Underground Storage Facilities

At the end of the mainline transmission system, and sometimes at its beginning and in between, underground natural gas storage and LNG (liquefied natural gas) facilities provide for inventory management, supply backup, and the access to natural gas to maintain the balance of the system. There are three principal types of underground storage sites used in the United States today: depleted reservoirs in oil and/or gas fields, aquifers, and salt cavern formations. In one or two cases mine caverns have been used. Two of the most important characteristics of an underground storage reservoir are the capability to hold natural gas for future use, and the rate at which natural gas inventory can be injected and withdrawn (its deliverability rate).

Most underground storage facilities, 327 out of 399 at the beginning of 2008, are depleted reservoirs, which are close to consumption centers and which were relatively easy to convert to storage service. In some areas, however, most notably the Midwestern United States, some natural aquifers have been converted to natural gas storage reservoirs. An aquifer is suitable for natural gas storage if the water-bearing sedimentary rock formation is overlaid with an impermeable cap rock. While the geology of aquifers is similar to that of depleted production fields, their use in natural gas storage usually requires more base (cushion) gas and greater monitoring of withdrawal and injection performance. Deliverability rates may be enhanced by the presence of an active water drive.

During the past 20 years, the number of salt cavern storage sites has grown significantly because of its rapid cycling (inventory turnover) capability coupled with its ability to respond to daily, even hourly, variations in customer needs. The large majority of salt cavern storage facilities have been developed in salt dome formations located in the Gulf Coast States. Salt caverns leached from bedded salt formations in Northeastern, Midwestern, and Western States have also been developed but the number has been limited due to a lack of suitable geology. Cavern construction is more costly than depleted field conversions when measured on the basis of dollars per thousand cubic feet of working gas capacity, but the ability to perform several withdrawal and injection cycles each year reduces the per-unit cost of each thousand cubic feet of natural gas injected and withdrawn.

Peak Shaving

Underground natural gas storage inventories provide suppliers with the means to meet peak customer requirements up to a point. Beyond that point the distribution system still must be capable of meeting customer short-term peaking and volatile swing demands that occur on a daily and even hourly basis. During periods of extreme usage, peaking facilities, as well as other sources of temporary storage, are relied upon to supplement system and underground storage supplies.

Peaking needs are met in several ways. Some underground storage sites are designed to provide peaking service, but most often LNG (liquefied natural gas) in storage and liquefied petroleum gas such as propane are vaporized and injected into the natural gas distribution system supply to meet instant requirements. Short-term linepacking is also used to meet anticipated surge requirements.

The use of peaking facilities, as well as underground storage, is essentially a risk-management calculation, known as peak-shaving. The cost of installing these facilities is such that the incremental cost per unit is expensive. However, the cost of a service interruption, as well as the cost to an industrial customer in lost production, may be much higher. In the case of underground storage, a suitable site may not be locally available. The only other alternative might be to build or reserve the needed additional capacity on the pipeline network. Each alternative entails a cost.

A local natural gas distribution company (LDC) relies on supplemental supply sources (underground storage, LNG, and propane) and uses linepacking to "shave" as much of the difference between the total maximum user requirements (on a peak day or shorter period) and the baseload customer requirements (the normal or average) daily usage. Each unit "shaved" represents less demand charges (for reserving pipeline capacity on the trunklines between supply and market areas) that the LDC must pay. The objective is to maintain sufficient local underground natural gas storage capacity and have in place additional supply sources such as LNG and propane air to meet large shifts in daily demand, thereby minimizing capacity reservation costs on the supplying pipeline.

www.ingramcontent.com/pod-product-compliance
Lightning Source LLC
Chambersburg PA
CBHW021036180526
45163CB00005B/2152